# Measuring Innovation
# &
# Technology Acceptance

Gordon C. Bruner II

# Table of Contents

# INTRODUCTION

It can be argued that societies benefit economically when new products are routinely introduced and readily adopted by consumers. It is no wonder then that the study of consumers' attitudes and behaviors with regard to innovations is a major topic in many academic disciplines. Research is also popular in industry because companies as well as other entities (governments, non-profit organizations) want people to accept new programs and offerings being introduced. Thus, a book that provides researchers with measures that have been used by social scientists in the study of adoption of innovations is ripe for development.

As one who has not only reviewed thousands of scales but also conducted studies having to do with innovation and technology acceptance, I thought it was time to use these experiences to develop a book that describes some measures that have been used by social scientists. There is no way to provide a review of all the potentially relevant measures in one bound book so the challenge was to select a small number of scales that would be useful to a wide group of researchers. The first decision was to focus on scales that had been reported in academic journals because the nature of measures and their usage can be more easily determined in those publications compared to research conducted in industry, the details of which are largely proprietary and unpublished. Further, the high standards that must be met by researchers who want their work considered for publication helps to raise the quality of the research that is ultimately published.

Specifically, the scales reviewed in this book have been reported in articles published in one or more journals closely related to consumer insight research: *Journal of Advertising, Journal of the Academy of Marketing Science, Journal of Consumer Research, Journal of Marketing, Journal of Marketing Research*, and *Journal of Retailing*. Hundreds of articles from those journals were examined, going back over 30 years. As might be imagined, there were hundreds of scales that were reported in that period relating to innovation and technology acceptance. Ultimately, thirty scales used in scholarly research were chosen for inclusion in this book. When there were multiple measures of the same constructs, I choose one I thought was "short and sweet," meaning I tried to balance high psychometric quality with few scale items.

There are two exceptions, however, to the rule of only providing one measure per construct: ease-of-use and usefulness. Those two constructs are so important, particularly in studies of technology acceptance, that two very different measures have been provided for each one.

For those researchers looking for measures of theoretical constructs not included in this book, they urged to refrain from rushing to create scales themselves. It is possible that they are out there somewhere. Admittedly, finding the ones most appropriate for one's purpose can be difficult, particularly for those without access to a good university library. Having said that, an easy starting point that anyone can use is with online scale databases. In the marketing arena, the massive database housed at MarketingScales.com is an obvious place to begin. Apart from that, a few other sources are worth checking out:

- Handbook of Management Scales:
  en.wikibooks.org/wiki/Handbook_of_Management_Scales

- International Personality Item Pool:
  ipip.ori.org

- ISWORLD Survey Instruments:
  people.ucalgary.ca/~newsted/list.htm

- Queendom:
  www.queendom.com/tests/index.htm

Finally, the table on the next page shows the layout of the scale reviews in this book. Each section of a review is described so that potential users can better understand what type of information is provided for the measures. For more explanation of the terminology employed in the reviews, see the Lexicon at:
https://www.marketingscales.com/research/lexicon.

# DESCRIPTION OF SCALE REVIEWS

**Scale Name**:

A short, descriptive title is assigned to each scale. Several issues are taken into account when assigning a title and the name may not be the one used researchers who reported using it.

**Scale Description**:

A few sentences are used to succinctly describe the theoretical construct being assessed. Also, the number of items, the number of points on the rating scale, and the response format (e.g., Likert, semantic differential) are typically stated, if known.

**Scale Origin**:

At least one usage of a scale is cited and, in some cases, there are several uses referenced. As for the source of a scale, information is provided, if known. However, in a substantial number of cases, a scale's origin and development is not known because the users did not identify it.

**Scale Reliability**:

For the most part, reliability is described in terms of internal consistency, most typically with Cronbach's alpha or construct reliability. Temporal stability (test-retest correlation) is provided in a few cases. With respect to those statistics, higher numbers are generally better. With particular regard to internal consistency, statistics below .60 if not .70 as well could be viewed as low reliability.

**Scale Validity**:

There are several types of validity and no one study is expected to fully "validate" a scale. While it is hoped that authors of each study would provide at least some evidence of a scale's validity, that is rarely the case. At the other extreme, some scale users have provided so much information in their articles that it is merely summarized in this section.

In those cases, readers are urged to consult the cited articles for more details.

**Comments**:

This field is used occasionally when the information does not fit well in the other sections.

**References**:

Every source cited in a review is referenced in this section. The six journals that were closely examined for articles with scales are the **Journal of Advertising, Journal of the Academy of Marketing Science, Journal of Consumer Research, Journal of Marketing, Journal of Marketing Research,** and **Journal of Retailing.** Citation of additional journals, books, proceedings, and other sources are provided when relevant.

**Scale Items**:

The statements, adjectives, or questions composing a scale are listed in this section. Also, an indication of the response format is provided unless it is has been stated in the **Scale Description** section. For example, if a measure is described as a "Likert-type" then it can be assumed that the extreme verbal anchors for the response scale were *strongly agree / strongly disagree* or some close variant. Where an item is followed by an (r) it means that the numerical response should be reverse coded when calculating scale scores. Other idiosyncrasies may be noted as well. For example, when slightly different versions of the same scale are discussed in the same review then an indication is given as to which items were used in particular studies.

# ADOPTION INTENTION

**Scale Description:**

This four-item, seven-point Likert-type scale is used to measure the inclination to buy a new product as soon as it is available. A sense of urgency to purchase the product earlier than others in one's relevant social groups is suggested in the items. There are direct and indirect versions of the scale. The difference between the two has to do with whether the items are responded to in the first person (direct version) or the third person (indirect version).

**Scale Origin:**

Fisher (1993) used the scale to measure students' intentions regarding a fictional new product idea: cordless headphones. Although not specifically stated, the scale is original (Fisher 1994). Items composing this scale were refined along with items intended to measure four other constructs. No information about this scale's development was provided. In general for the scales in the study, item-to-total correlations had to be higher than .50 and items had to load on hypothesized factors. Items that did not fit these criteria were eliminated before use in the main study.

**Scale Reliability:**

A Cronbach's alpha of .93 was reported for the scale by Fisher (1993; Fisher and Price 1992) for the group receiving direct questioning. For the group receiving indirect questioning, the alpha was .81 (Fisher 1993).

**Scale Validity:**

The validity of the scale was not specifically addressed by Fisher (1993; Fisher and Price 1992). However, Fisher and Price (1992) did state that the variance extracted for the scale was .79.

**References:**

Fisher, Robert J. (1994), personal correspondence.

Fisher, Robert J. (1993), "Social Desirability Bias and the Validity of Indirect Questioning," *Journal of Consumer Research,* 20 (September), 303-15.

Fisher, Robert J. and Linda L. Price (1992), "An Investigation into the Social Context of Early Adoption Behavior," *Journal of Consumer Research,* 19 (December), 477-86.

**Scale Items:**[1]

Direct version of scale

1. I would like to buy a _____ today, if possible.
2. I will try to buy one of the products as soon as I can.
3. I am likely to be one of the first _____ to buy a _____.[2]
4. I will probably purchase one of the new products soon after they are on the market.

Indirect version of scale

The typical student will . . .

1. . . . want to buy a _____ today, if possible.
2. . . . try to buy one of the products as soon as s/he can.
3. . . . want to be among the first _____ to buy a _____.[2]
4. . . . probably purchase one of the new products soon after they are on the market.

---

[1] The items were supplied by Fisher (1994). The generic name of the product should be placed in the blanks.

[2] The first blank in this item should be filled with a name for an important reference group of which the respondents are members, e.g., employees, students. If a general term such as *persons* is used, it changes the meaning of the scale somewhat.

# ATTITUDE TOWARD THE INNOVATION

**Scale Description:**

This eleven item, seven-point scale measures the attitude a consumer has toward a specific new product. The scale is broad enough to tap not only into what a person thinks about a specified product, but also how it is thought others might respond to it. The scale was referred to as *attractiveness of innovation* by Boyd and Mason (1999).

**Scale Origin:**

The scale was constructed by Boyd and Mason (1999). A set of potential items were developed based on Rogers' (1962) list of innovation characteristics as well as feedback from a focus group. Several judges examined the items for face validity and the scale was tested on a convenience sample (n = 55 college students).

**Scale Reliability:**

An alpha of .91 was reported for the scale (Boyd and Mason 1999).

**Scale Validity:**

No examination of the scale's validity was reported by Boyd and Mason (1999). However, they did state that an exploratory factor analysis showed the items to be unidimensional.

**References:**

Boyd, Thomas C. and Charlotte H. Mason (1999), "The Link Between Attractiveness of 'Extrabrand' Attributes and the Adoption of Innovations," *Journal of the Academy of Marketing Science*, 27 (3), 306-319.

Rogers, Everett M. (1962), *Diffusion of Innovation*, New York: Free Press.

## Scale Items:[1]

1. _____ is a great idea.
2. _____ would be fun to own.
3. This is the best way to improve the quality of _____.
4. Many people will buy _____.
5. _____ is here to stay.
6. _____ fills a real need for me.
7. _____ is a big improvement over existing _____.
8. _____ can give me real value.
9. _____ is just another gimmick. (r)
10. _____ fills a need for many people.
11. Many people will believe _____ is worth the cost.

---

[1] The actual product name should be placed in the blanks. The name of the product category should be in the blank of #4 and the second blank in #7. Response scale anchors ranged from *do not agree at all* (1) to *completely agree* (7).

# ATTITUDE TOWARD THE PRODUCT (HIGH TECH)

**Scale Description:**

This semantic differential scale is used to measure a person's general evaluation of a high tech good or service.

**Scale Origin:**

Although many of the items in this scale have been used previously, particularly in the measurement of attitude toward the brand, this version by Roehm and Sternthal (2001) appears to be unique. Not only are there several bi-polar adjectives that as a set have not had much previous usage but the scale as a whole was meant to be used with high tech products rather than any other goods or services.

**Scale Reliability:**

Three versions of the scale were used by Roehm and Sternthal (2001) in the four experiments they conducted. Two slightly different ten-item versions were used that had alphas of .93 (Study 1) and .90 (Study 3). The alphas for the eight- and 13-item versions were .90 (Study 4) and .93 (Study 2), respectively.

**Scale Validity:**

No examination of the scale's validity was reported by Roehm and Sternthal (2001). However, they did state that factor analysis indicated the various versions of the scale were unidimensional.

**Reference:**

Roehm, Michelle L. and Brian Sternthal (2001), "The Moderating Effect of Knowledge and Resources on the Persuasive Impact of Analogies," *Journal of Consumer Research*, 28 (September), 257-272.

## Scale Items:[1]

1. like / dislike
2. useful / not useful
3. high-tech / not high-tech
4. good / bad
5. high quality / low quality
6. practical / impractical
7. worth owning / not worth owning
8. impressive / not impressive
9. valuable / not valuable
10. advanced / not advanced
11. dependable / not dependable
12. simple / complex
13. convenient / inconvenient

---

[1] The eight-item version of the scale was composed of #1, #2, #4, #6-#10. The ten-item version was composed of items #1-#10 and the 13-item version used all of the listed items (Roehm and Sternthal 2001). A seven-point response format appears to have been used.

# BEHAVIORAL CONTROL

## Scale Description:

The extent to which a person feels that he/she is in control of some object or process is measured in this scale with four, seven-point Likert-type items. Collier and Sherrell (2010) used the scale with a self-service technology (SST) but it appears to be amenable for use in a wider context.

## Scale Origin:

The scale used by Collier and Sherrell (2010) is original but is based on key words and concepts from measures used by Dabholkar (1996), Yen, and Gwinner (2003), as well as Zhu (2002). The authors pretested the scale along with the other scales in their questionnaire with 500 patrons of a company. Although psychometric information was not provided about this scale based on the pretest results, it was stated in general that all items loaded on their respective constructs and each scale had a Cronbach's alpha greater than or equal to .70. Analyses of the main study's results were based on responses from 2,246 customers who had used a company's SST.

## Scale Reliability:

Based on the main study conducted by Collier and Sherrell (2010), the scale's alpha was .901.

## Scale Validity:

Using CFA, Collier and Sherrell (2010) concluded that their measurement model fit the data. They also found support for the convergent and discriminant validities of this scale as well as the others in their study. The AVE for this scale was .74.

## References:

Collier, Joel E. and Daniel L. Sherrell (2010), "Examining the Influence of Control and Convenience in a Self-service Setting," *Journal of the Academy of Marketing Science*, 38 (4), 490-509.

Dabholkar, Pratibha A. (1996), "Consumer Evaluations of New Technology-Based Self-Service Options: An Investigation of Alternative Models of Service Quality," *International Journal of Research in Marketing*, 13 (1), 29-51.

Yen, Hsiu Ju Rebecca and Kevin P. Gwinner (2003), "Internet Retail Customer Loyalty: The Mediating Role of Relational Benefits," *International Journal of Service Industry Management*, 14 (5), 483–500.

Zhu, Zhen (2002), *Fix It or Leave It: Antecedents and Consequences of Perceived Control in Technology-based Self-Service Failure Encounters*, doctoral dissertation, University of Illinois at Chicago: Chicago.

## Scale Items:[1]

1. I feel in control using _____.
2. _____ lets the customer be in charge.
3. While using _____, I feel decisive.
4. _____ gives me more control over _____.

---

[1] The short space in the items should be filled with a name for the focal object the respondents have interacted with. The long space in #4 should be filled with a more specific activity, e.g., streaming movies.

# BEHAVIORAL INTENTION (GENERAL)

**Scale Description:**

Using semantic differentials, this scale measures the expressed inclination of a person to engage in a specified behavior. In the studies cited below, the behavior was a purchase in the context of technology or new products. Paired with the proper instructions, the scale items are general enough to refer to behaviors that are not explicitly purchases if desired (e.g., likelihood of trying or using an innovation).

**Scale Origin:**

The source of the scale is not clear. Work by Fishbein (Fishbein and Ajzen 1975; Ajzen and Fishbein 1980) is a possibility although only item #1 (below) figures prominently in his books as a way to measure behavioral intention. Among the first uses of the multi-item scale in a consumer context was with respect to purchase of a new diet suppressant (Oliver and Bearden 1985).

Ko, Cho, and Roberts (2005) gathered data for their experiment in both the U.S. and Korea. In doing that, the scales for the Korean version were developed using a back-translation process. Contact those authors directly for the Korean version.

**Scale Reliability:**

The four item version of the scale used by Oliver and Bearden (1985) had a construct reliability of .87. The alphas that have been reported for the three item version of the scale are .91 (Bruner and Kumar 2000), .89 (Ko, Cho, and Roberts 2005), and .88 (MacKenzie and Spreng 1992).

**Scale Validity:**

In none of the studies was the scale's validity fully addressed. Despite that, some evidence of the scale's validity is available. Oliver and Bearden (1985) used CFA and found that their measurement model fit the data well. The variance extracted for behavioral intention was .67.

Ko, Cho, and Roberts (2005) reported their model fit their data well which, indirectly, provides some limited evidence of the scale's validity.

Although not specifically examining the validity of behavioral intention, a correlation matrix was provided by MacKenzie and Spreng (1992) between the items in the behavioral intention scale as well as several others. The inter-correlations of the behavioral intention items ranged between .47 and .88 which provides some evidence that the items are measuring the same thing. In contrast, the correlations between the intention items and those measuring related but theoretically distinct constructs were much lower.

**Comments:**

This is a greatly condensed description of the scale's usage compared to what is provided in the *Marketing Scales Handbooks* (e.g., Bruner 2009, 2013). As noted in those books, this measure and its variations make it one of the most used scales in scholarly consumer insight research. No doubt, that is due to the construct's extreme importance in understanding behavior as well as the scale's adaptability for many different contexts.

**References:**

Ajzen, Icek and Martin Fishbein (1980), *Understanding Attitudes and Predicting Social Behavior*, Englewood Cliffs, NJ: Prentice-Hall Inc.

Bruner II, Gordon C. (2009), *Marketing Scales Handbook: A Compilation of Multi-Item Measures for Consumer Behavior & Advertising* (Volume 5), Carbondale, IL: GCBII Productions.

Bruner II, Gordon C. (2013), *Marketing Scales Handbook: The Top 20 Multi-Item Measures Used in Consumer Research*, CreateSpace Independent Publishing Platform.

Bruner II, Gordon C. and Anand Kumar (2000), "Web Commercials and Advertising Hierarchy-of-Effects," *Journal of Advertising Research*, 40 (Jan-Apr), 35-42.

Fishbein, Martin and Icek Ajzen (1975), *Belief, Attitude, Intention, and Behavior: An Introduction to Theory and Research*, Reading, Mass.: Addison-Wesley.

Ko, Hanjun, Chang-Hoan Cho, and Marilyn S. Roberts (2005), "Internet Uses and Gratifications: A Structural Equation Model of Interactive Advertising," *Journal of Advertising*, 34 (2), 57-70.

MacKenzie, Scott B. and Richard A. Spreng (1992), "How Does Motivation Moderate the Impact of Central and Peripheral Processing on Brand Attitudes and Intentions," *Journal of Consumer Research*, 18 (March), 519-529.

Oliver, Richard L. and William O. Bearden (1985), "Crossover Effects in the Theory of Reasoned Action: A Moderating Influence Attempt," *Journal of Consumer Research*, 12 (December), 324-340.

## Scale Items:[1]

1. unlikely / likely
2. improbable / probable
3. impossible / possible
4. uncertain / certain

---

[1] Oliver and Bearden (1985) used all four items whereas Bruner and Kumar (2000), Ko, Cho, and Roberts (2005), Lacher and Mizerski (1994), and MacKenzie and Spreng (1992) used the first three. Most if not all of the studies used a seven-point response format.

# CHANGE SEEKING

## Scale Description:

The scale is composed of seven statements measuring the degree to which a person expresses a desire for variation or stimulation in his/her life. The scale can be viewed as a measure of optimum stimulation level (e.g., Campbell and Goodstein 2001) or inherent novelty seeking (Dabholkar and Bagozzi 2002).

## Scale Origin:

The scale used by Campbell and Goodstein (2001) came from Steenkamp and Baumgartner (1995). The latter developed it to be a short form of the CSI (Change Seeker Index), the 95 item instrument created by Garlington and Shimota (1964). The studies conducted by Steenkamp and Baumgartner (1995) first reduced the scale from 95 items to seven and then cross-validated those seven in three countries and with two types of subjects.

Interestingly, six of the items in this short form are also a subset of the well-known 40 item arousal seeking scale by Mehrabian and Russell (1974). That is the source cited for the version of the scale used by Dabholkar and Bagozzi (2002).

## Scale Reliability:

Baumgartner and Steenkamp (2001) reported an overall alpha of .75 in their pan-European survey with alphas for individual countries ranging from .60 to .81. In Study 3 by Campbell and Goodstein (2001), the alpha of the scale was .82. An alpha of .72 was reported for the version of the scale used by Dabholkar and Bagozzi (2002).

## Scale Validity:

The purpose of the study by Baumgartner and Steenkamp (2001) was to examine response styles as a source of contamination in questionnaire measures and the effect that might have on validity of conclusions drawn from such data. Although most of the results were reported at an overall level, one finding pertinent to this scale was that the mean level of contamination in scale scores was estimated to be 2% (ranging from 1%-4% for eleven European countries), among the lowest average amounts of contamination found for the 14 scales that were examined.

Evidence was provided by Dabholkar and Bagozzi (2002) in support of the convergent and discriminant validity for the version of the scale they used.

**References:**

Baumgartner, Hans and Jan-Benedict E.M. Steenkamp (2001), "Response Styles in Marketing Research: A Cross-National Investigation," *Journal of Marketing Research*, 38 (May), 143-156.

Campbell, Margaret C. and Ronald C. Goodstein (2001), "The Moderating Effect of Perceived Risk on Consumers' Evaluations of Product Incongruity: Preference for the Norm," *Journal of Consumer Research*, 28 (December), 439-449.

Dabholkar, Pratibha and Richard P. Bagozzi (2002), "An Attitudinal Model of Technology-Based Self-Service: Moderating Effects of Consumer Traits and Situational Factors," *Journal of the Academy of Marketing Science*, 30 (3), 184-201.

Garlington, Warren K. and Helen E. Shimota (1964), "The Change Seeker Index: A Measure of the Need for Variable Stimulus Input," *Psychological Reports*, 14, 919-924.

Mehrabian, Albert and James A. Russell (1974), *An Approach to Environmental Psychology*, Cambridge, MA: The MIT Press.

Steenkamp, Jan-Benedict E.M. and Hans Baumgartner (1995), "Development and Cross-Cultural Validation of a Short form of CSI as a Measure of Optimum Stimulation Level," *International Journal of Research in Marketing*, 12, 97-104.

## Scales Items:[1]

1. I like to continue doing the same old things rather than trying new and different things. (r)
2. I like to experience novelty and change in my daily routine.
3. I like a job that offers change, variety, and travel, even if it involves some danger.
4. I am continually seeking new ideas and experiences.
5. I like continually changing activities.
6. When things get boring, I like to find some new and unfamiliar experience.
7. I prefer a routine way of life to an unpredictable one full of change. (r)

---

[1] The response format used by Campbell and Goodstein (2001) was not described. Steenkamp and Baumgartner (1995) used a five-point scale ranging from *completely false* to *completely true*. The final version of the scale used by Dabholkar and Bagozzi (2002) was composed of items the same or very similar to #2, #4-#6 and a seven-point Likert-type response format.

# COMMUNICATION OF PRODUCT USAGE OUTCOMES

**Scale Description:**

With three, five-point Likert-type items, this scale measures the degree to which a consumer understands the consequences of using a product and can communicate them to others. The scale may be better suited for a product with benefits of a functional nature as opposed to those that are hedonic or social.

**Scale Origin:**

Van Ittersum and Feinberg (2010) did not identify the source of the scale except to say they "relied on published, validated scales" in three of their studies. The authors called the scale *perceived observability,* referring to a construct long identified by Rogers (e.g., 2003) as one of the key characteristics of successful innovations. In Study 1, 212 superintendents of U.S. golf courses were surveyed about their intentions to adopt an advanced grass mower. In Study 2, 266 U.S. farm operators were asked about their intentions to adopt an autoguidance system for their tractors. Finally, 354 U.S. college students were asked in Study 3 about their intentions to adopt cell phones with GPS.

**Scale Reliability:**

Alphas for the scale were .71, .64, and .83 in Studies 1, 2, and 3, respectively, by Van Ittersum and Feinberg (2010).

**Scale Validity:**

Van Ittersum and Feinberg (2010) did not address the scale's validity.

**References:**

Rogers, Everett M. (2003), *Diffusion of Innovations*, New York: The Free Press.

Van Ittersum, Koert and Fred M. Feinberg (2010), "Cumulative Timed Intent: A New Predictive Tool for Technology Adoption," *Journal of Marketing Research*, 47 (5), 808-822.

**Scale Items:**[1]

1. I have no difficulty telling others about the results of using
   _____.

2. I believe I could communicate to others the consequences of using
   _____.

3. The results of using _____ are apparent to me.

---

[1] The name of the product/brand should be placed in the blanks.

# COMPATIBILITY OF THE PRODUCT

**Scale Description:**

Three, seven-point Likert-type statements are used to measure the degree to which a consumer believes that a good or service is well-suited to his/her needs and lifestyle. Because this is one of the five key characteristics that are thought to influence adoption of innovations (Rogers 2003), this construct is most typically examined with respect to new products rather than mature ones.

**Scale Origin:**

The scale was adapted by Meuter et al. (2005) from key phrases and concepts in a scale by Moore and Benbasat (1991).

**Scale Reliability:**

Alphas of .95 and .97 were reported by Meuter et al. (2005) for use of the scale in Studies 1 and 2, respectively.

**Scale Validity:**

At a general level, Meuter et al. (2005) tested a measurement model containing all of their constructs and indicators. Its fit was acceptable. The factor loadings were reported to be significant and evidence of discriminant validity was provided for each construct using two different tests (confidence interval, variance extracted).

**Comments:**

See a variation on this scale used by Van Ittersum and Feinberg (2010).

**References:**

Meuter, Matthew L., Mary Jo Bitner, Amy L. Ostrom, and Stephen W. Brown (2005), "Choosing Among Alternative Service Delivery Modes: An Investigation of Customer Trial of Self-Service Technologies," *Journal of Marketing*, 69 (April), 61-83.

Moore, Gary C. and Izak Benbasat (1991), "Development of an Instrument to Measure the Perceptions of Adopting an Information Technology Innovation," *Information Systems Research*, 2 (3), 192-223.

Rogers, Everett M. (2003), *Diffusion of Innovations*, New York: The Free Press.

Van Ittersum, Koert and Fred M. Feinberg (2010), "Cumulative Timed Intent: A New Predictive Tool for Technology Adoption," *Journal of Marketing Research*, 47 (5), 808-822.

**Scale Items:**[1]

1. Using the _____ is compatible with my lifestyle.
2. Using the _____ is completely compatible with my needs.
3. The _____ fits well with the way I like to get things done.

---

[1] The name of the good or service should be placed in the blanks.

# COMPLEXITY OF THE INNOVATION

**Scale Description:**

Three, seven-point items are used to measure the degree of challenge a consumer perceives there to be in learning to use a new good or service.

**Scale Origin:**

The source of the scale used by Wood and Moreau (2006) was not stated but appears to have been developed by them.

**Scale Reliability:**

The scale was administered by Wood and Moreau (2006) to participants at two points in time (if not three) and the alphas were reported to be over .90 each time.

**Scale Validity:**

No information regarding the scale's validity was discussed by Wood and Moreau (2006).

**Reference:**

Wood, Stacy L. and C. Page Moreau (2006), "From Fear to Loathing? How Emotion Influences the Evaluation and Early Use of Innovations," *Journal of Marketing*, 70 (3), 44-57.

**Scale Items:**[1]

1. How difficult do you expect/perceive the _____ to be to use?[2]
2. How long would it take to learn to use the _____?[3]
3. How much of a challenge is there in using the _____?[4]

[1] As shown in #1, the phrasing of the items might need to change somewhat if administered before use of the innovation (expectations) vs. when participants have already used it (perceptions).

[2] The verbal anchors were not described by Wood and Moreau (2006) but may have been something like *very simple / very difficult* for this statement.

[3] The verbal anchors were not described by Wood and Moreau (2006) but may have been something like *very little time / a lot of time* for this statement.

[4] The verbal anchors were not described by Wood and Moreau (2006) but may have been something like *very little challenge / very challenging* for this statement.

# EASE OF USE (GENERAL)

**Scale Description:**

The seven point semantic differential scale measures a person's beliefs concerning the time and effort involved in a specified activity. The activity examined by Dabholkar (1994) was ordering in a fast-food restaurant and two options were compared: touch screen ordering versus verbally placing the order with an employee. Dabholkar and Bagozzi (2002) just examined the touch screen option.

**Scale Origin:**

The origin of the scale appears to be Dabholkar (1994). Refinement of the scale occurred with a pretest sample that consisted of 141 undergraduate students. The scale had alphas of .88 (touch screen ordering) and .80 (verbal ordering).

**Scale Reliability:**

Dabholkar (1994) reported construct reliabilities of .92 and .86 for the touch screen and verbal versions of the scale, respectively. An alpha of .90 was reported for the version of the scale used by Dabholkar and Bagozzi (2002).

**Scale Validity:**

Results of confirmatory and exploratory factor analyses provided by Dabholkar (1994) indicated that both versions of the scale were unidimensional. Evidence was provided by Dabholkar and Bagozzi (2002) in support of the scale's convergent and discriminant validity.

**Comments:**

See also Van Dolen, Dabholkar, and Ruyter (2007) for a modified version of this scale where the items were rephrased as statements for use with a Likert-type response format.

## References:

Dabholkar, Pratibha (1994), "Incorporating Choice into an Attitudinal Framework: Analyzing Models of Mental Comparison Processes," *Journal of Consumer Research*, 21 (June), 100-118.

Dabholkar, Pratibha and Richard P. Bagozzi (2002), "An Attitudinal Model of Technology-Based Self-Service: Moderating Effects of Consumer Traits and Situational Factors," *Journal of the Academy of Marketing Science*, 30 (3), 184-201.

Van Dolen, Willemijn M., Pratibha A. Dabholkar, and Ko de Ruyter (2007), "Satisfaction with Online Commercial Group Chat: The Influence of Perceived Technology Attributes, Chat Group Characteristics, and Advisor Communication Style," *Journal of Retailing*, 83 (3), 339-358.

## Scale Items:[1]

1. will be complicated / will be simple
2. will take a lot of effort / will take a little effort
3. will be confusing / will be clear
4. will take a long time / will take a short time
5. will require a lot of work / will require little work
6. will be slow / will be fast

---

[1] The scale stem used by Dabholkar (1994; Dabholkar and Bagozzi 2002) was "Using a _____ to order fast food . . . " and "touch screen" or "verbal" were placed in the blank. Only one anchor for each pair was explicitly stated in the articles; the others are hypothetical reconstructions. Dabholkar (1994) used all six items while the final version of the scale used by Dabholkar and Bagozzi (2002) was composed of items #2, #4-#6.

# EASE OF USE OF THE PRODUCT

**Scale Description:**

The scale has three, seven-point Likert-type statements that measure the degree to which a consumer believes a good or service to be free from effort when being used. Meuter et al. (2005) referred to this scale as *complexity* because they were studying the five key characteristics thought to influence adoption of innovations (Rogers 2003).

**Scale Origin:**

Although Meuter et al. (2005) drew from a scale by Moore and Benbasat (1991), the key phrases are quite common among the variety of ease-of-use scales that have been created at least since Davis (1989).

**Scale Reliability:**

Alphas of .83 and .88 were reported by Meuter et al. (2005) for use of the scale in Studies 1 and 2, respectively.

**Scale Validity:**

At a general level, Meuter et al. (2005) tested a measurement model containing all of their constructs and indicators. Its fit was acceptable. The factor loadings were reported to be significant and evidence of discriminant validity was provided for each construct using two different tests (confidence interval, variance extracted).

**References:**

Davis Fred D. (1989), "Perceived Usefulness, Perceived Ease of Use and User Acceptance of Information Technology," *MIS Quarterly*, 13 (2), 319-339.

Meuter, Matthew L., Mary Jo Bitner, Amy L. Ostrom, and Stephen W. Brown (2005), "Choosing Among Alternative Service Delivery Modes: An Investigation of Customer Trial of Self-Service Technologies," *Journal of Marketing*, 69 (April), 61-83.

Moore, Gary C. and Izak Benbasat (1991), "Development of an Instrument to Measure the Perceptions of Adopting an Information Technology Innovation," *Information Systems Research*, 2 (3), 192-223.

Rogers, Everett M. (2003), *Diffusion of Innovations*, New York: The Free Press.

**Scale Items:**[1]

1. I believe that the _____ is cumbersome to use. (r)
2. It is difficult to use the _____. (r)
3. I believe that the _____ is easy to use.

---

[1] The name of the good or service should be placed in the blanks.

# EMERGENT NATURE

**Scale Description:**

A person's ability to imagine how new product concepts could be developed in order to be more useful and relevant to consumers is measured in this scale with eight, seven-point Likert-type items.

**Scale Origin:**

Hoffman, Kopalle, and Novak (2010) conceived of, constructed, and validated the scale in a well-executed series of studies. The authors viewed the construct as resulting from a mixture of personality traits and processing abilities. Indeed, many of those traits and abilities were incorporated into the validation process. (See the article's web appendix for more details of the studies, samples, and scales.)

Based upon analysis of pretest data, a large set of preliminary items was reduced to 17 which were further purified and validated in Study 1. Analysis of that study's data was based on responses from 1124 adults living in 21 different countries whose native language was English. In Studies 2 and 3, the emergent nature scale was used to test the some predictions which provided more evidence of the scale's validity.

**Scale Reliability:**

Per Study 1, the scale's alpha was .93 (Hoffman, Kopalle, and Novak 2010).

**Scale Validity:**

A variety of evidence was provided by Hoffman, Kopalle, and Novak (2010) in support of the scale's unidimensionality and discriminant validity. Evidence of the scale's predictive validity came from Studies 2 and 3.

## Comments:

With data from the Study 1 sample, the scale's mean was calculated to be 36.99 and the standard deviation was 9.78 (Hoffman, Kopalle, and Novak 2010, web appendix).

## Reference:

Hoffman, Donna L., Praveen K. Kopalle, and Thomas P. Novak (2010), "The 'Right' Consumers for Better Concepts: Identifying Consumers High in Emergent Nature to Develop New Product Concepts," *Journal of Marketing Research*, 47 (5), 854-865.

## Scale Items:

1. When I hear about a new product or service idea, it is easy to imagine how it might be developed into an actual product or service.
2. Even if I don't see an immediate use for a new product or service, I like to think about how I might use it in the future.
3. When I see a new product or service idea, it is easy to visualize how it might fit into the life of an average person in the future.
4. If someone gave me a new product or service idea with no clear application, I could "fill in the blanks" so someone else would know what to do with it.
5. Even if I don't see an immediate use for a new product or service, I like to imagine how people in general might use it in the future.
6. I like to experiment with new ideas for how to use products and services.
7. I like to find patterns in complexity.
8. I can picture how products and services of today could be improved to make them more appealing to the average person.

# GADGET LOVING

**Scale Description:**

The scale is composed of eight, seven-point Likert-type statements that measure the degree to which a consumer expresses high intrinsic motivation to adopt and use innovative, technology-based goods and services. Those scoring high on the scale are referred to as *gadget lovers*.

**Scale Origin:**

The scale is original to Bruner and Kumar (2006, 2007). The scale was developed in a series of studies, some of the details of which are described in the published article (Bruner and Kumar 2007) but much more is in an unpublished paper (Bruner and Kumar 2006).

**Scale Reliability:**

Among the several times the scale was used, the alphas ranged from .89 (Study 4, 188 college students) to .94 (Study 3, 1,366 customers of a wireless provider). The temporal stability (three month test-retest) of the scale was examined in Study 4 with 71 students who had also completed the scale three months earlier. The test-retest correlation was .74.

**Scale Validity:**

Quite a bit of information bearing on the scale's validity is provided in the published version of the article (Bruner and Kumar 2007) and more is in the unpublished version (Bruner and Kumar 2006). In brief, support was provided for the scale's content, convergent, discriminant, and concurrent validities. One concern about the scale which the authors mentioned had to do with item #6 (below). It was weaker than the other items and is a candidate for elimination. To maintain the scale's content validity, the authors suggested replacing the item with something referring more generally to ongoing search activity, e.g., *reading about new gadgets soon to be released is something I enjoy doing*.

**Comments:**

See also Shoham and Pesämaa (2013) who re-examined the scale and not only found support for its psychometric quality but its cross-cultural generalizability as well.

**References:**

Bruner II, Gordon C. and Anand Kumar (2006), "Gadget Lovers," *Office of Scale Research Technical Report #0602*, scaleresearch.siuc.edu/tr0602.pdf.

Bruner II, Gordon C., Anand Kumar (2007), "Gadget Lovers," *Journal of the Academy of Marketing Science*, 35 (3), 329-339.

Shoham, Aviv and Ossi Pesämaa (2013), "Gadget Loving: A Test of an Integrative Model," *Psychology & Marketing*, 30 (3), 247-262.

**Scale Items:**

1. Despite their age, I love to play around with technological gadgets.
2. Even if they aren't the newest things on the market, learning how to operate technological products is interesting to me.
3. Old or new, playing with technological products brings me a lot of enjoyment.
4. Others may not understand it but it's kind of a thrill to play with products that have a high-tech component.
5. If I was alone for several hours I could entertain myself easily if I had lots of gadgets to play with.
6. Leafing through catalogs from high-tech vendors such as Sharper Image and Dell is something I like to do.
7. It is easy for me to spend a lot of time playing around with almost any kind of technological device.
8. Some people find it irritating but I enjoy figuring out how to get technological goods and services to work.

# INNOVATION ABILITY OF THE COMPANY

**Scale Description:**

With three, seven-point Likert-type items, this scale measures a person's belief that a company is capable of creating original and interesting new products.

**Scale Origin:**

The scale was used by Schreier, Fuchs, and Dahl (2012) in Study 3 of the four studies discussed in their article. The sample was composed of 466 consumers recruited from a market research agency. The authors said that they adapted the scale from work by Rindfleisch and Moorman (2001). Since no scale in that piece bears any similarity to this innovation ability scale, it is best to consider Schreier, Fuchs, and Dahl (2012) as the source of the scale.

**Scale Reliability:**

The alpha for the scale was .92 (Schreier, Fuchs, and Dahl 2012, p. 27).

**Scale Validity:**

Schreier, Fuchs, and Dahl (2012) did not discuss the scale's validity, at least, they did not comment on it directly. In Study 3, the authors not only used this scale to measure innovative ability but also another scale that they used in each of the four studies. The two measures were highly correlated and, in fact, a CFA showed there was a single factor. That provides some evidence of the scale's convergent validity.

**References:**

Rindfleisch, Aric and Christine Moorman (2001), "The Acquisition and Utilization of Information in New Product Alliances: A Strength-of-Ties Perspective," *Journal of Marketing*, 65 (April), 1–18.

Schreier, Martin, Christoph Fuchs, and Darren W. Dahl (2012), "The Innovation Effect of User Design: Exploring Consumers' Innovation Perceptions of Firms Selling Products Designed by Users," *Journal of Marketing*, 76 (5), 18–32.

**Scale Items:**

1. I think the firm has the ability to develop really innovative new products.
2. The firm is in the position to derive very original product ideas.
3. The company has a large potential to foster creativity.
4. I think the firm can create very interesting new products.

# INNOVATIVENESS IMPORTANCE TO THE CONSUMER (TECHNOLOGICAL)

**Scale Description:**

The level of importance a consumer places on knowing about and owning new technological products is measured in this six item, seven-point Likert-type scale.

**Scale Origin:**

Shalev and Morwitz (2012) appear to have created the scale based on inspiration received from items in scales by Goldsmith and Hofacker (1991) as well as Martínez and Montaner (2006). The scale was used by Shalev and Morwitz (2012) in Study 2 (n = 77 members of an online panel) but the development of the scale was not described.

**Scale Reliability:**

The scale's alpha was .93 (Shalev and Morwitz 2012, p. 969).

**Scale Validity:**

The scale's validity was not discussed by Shalev and Morwitz (2012).

**References:**

Goldsmith, R. E. and C. F. Hofacker (1991), "Measuring Consumer Innovativeness," *Journal of the Academy of Marketing Science*, 19 (3), 209–21.

Martínez, Eva and Teresa Montaner (2006), "The Effect of Consumer's Psychographic Variables upon Deal-Proneness," *Journal of Retailing and Consumer Services*, 13 (3), 157–68.

Shalev, Edith and Vicki G. Morwitz (2012), "Influence via Comparison-Driven Self-Evaluation and Restoration: The Case of the Low-Status Influencer," *Journal of Consumer Research*, 38 (5), 964-980.

**Scale Items:**

1. It is important for me to try new and different technological products.
2. Compared to my friends, I own many technological products.
3. It is important for me to keep up with contemporary technologies.
4. It is important for me to be among the first people to own new technological products.
5. It is important for me to be able to make recommendations to others about technological products.
6. When I see a technological product somewhat different from the usual, it is important for me to check it out.

---

# INNOVATIVENESS OF THE CONSUMER (PRODUCT SPECIFIC)

## Scale Description:

This six-item, five-point Likert-type scale is used to measure the tendency to learn about and adopt innovations (new products) within a specific domain of interest. The scale is intended to be distinct from a generalized personality trait at one extreme and a highly specific, single product purchase at the other extreme.

## Scale Origin:

The scale was constructed by Goldsmith and Hofacker (1991). Their desire was to develop a short, flexible measure of consumer innovativeness modeled after King and Summers's (1970) measure of opinion leadership.

## Scale Reliability:

An alpha of .83 was calculated for the scale in Study 3 (fashion) by Flynn, Goldsmith, and Eastman (1996; Goldsmith 1997). Because they used multiple studies, several examinations of the scale's internal consistency were reported by Goldsmith and Hofacker (1991). The following alphas were calculated: .83 (Study 2, records); .82 (Study 3, records); .79 (Study 4, fashion); .81 (Study 4, electronics); .88 (Study 5, records, n = 75); .90 (Study 4, records, n = 70); and .85, .83, and .83 for records, fashion, and scent, respectively, in Study 6. The temporal stability of the scale as measured in Study 5 was .86.

## Scale Validity:

The validity of the scale was not directly examined by Flynn, Goldsmith, and Eastman (1996). However, because it was part of an effort to examine the nomological validity of two other scales, some sense of its own nomological validity can be gained. For example, high positive correlations were found between innovativeness and opinion leadership, product involvement, and product knowledge. No relationship was found between innovativeness and opinion seeking.

The validity of the scale was examined in detail by Goldsmith and Hofacker (1991). Data from Study 2 were subjected to both exploratory and confirmatory factor analysis with similar results: the items loaded on the same factor and were unidimensional. Also in Study 2, a pattern of significant positive correlations with seven criterion measures provided evidence of the scale's criterion validity. Studies 3, 4, and 5 had results very similar to Study 2's in that the evidence supported the notion that the scale was unidimensional and had criterion validity. In addition, Study 5 provided evidence of discriminant validity in that the scale was not significantly correlated with yea-saying or social desirability bias. Study 6 measured innovativeness in three product categories (music, fashion, and scent). Using a multi-trait, multi-method approach, the authors concluded that there was strong evidence of convergent and discriminant validity.

**Comments:**

The authors believe that the scale is most appropriate for studying products in categories that are purchased rather frequently. Products that are bought less often might not be measured as well with this scale because there are fewer attitudes and behaviors for consumers to draw on in making their responses.

For other uses of the scale, see Goldsmith and Flynn (1992) as well as Flynn and Goldsmith (1993). German and French versions can be found in Goldsmith, d'Hauteville, and Flynn (1998).

**References:**

Flynn, Leisa R. and Ronald E. Goldsmith (1993), "Identifying Innovators in Consumer Service Markets," *The Service Industries Journal*, 13 (July), 97–109.

Flynn, Leisa R., Ronald E. Goldsmith, and Jacqueline K. Eastman (1996), "Opinion Leaders and Opinion Seekers: Two New Measurement Scales," *Journal of the Academy of Marketing Science*, 24 (Spring), 137–47.

Goldsmith, Ronald E. (1997), personal correspondence.

Goldsmith, Ronald E., Francois d'Hauteville, and Leisa R. Flynn (1998), "Theory and Measurement of Consumer Innovativeness," *European Journal of Marketing*, 32 (3/4), 340–53.

Goldsmith, Ronald E. and Leisa Reinecke Flynn (1992), "Identifying Innovators in Consumer Markets," *European Journal of Marketing*, 26 (12), 42–55.

Goldsmith, Ronald E. and Charles F. Hofacker (1991), "Measuring Consumer Innovativeness," *Journal of the Academy of Marketing Science*, 19 (Summer), 209–21.

King, Charles W. and John O. Summers (1970), "Overlap of Opinion Leadership Across Consumer Product Categories," *Journal of Marketing Research*, 7 (February), 43–50.

**Scale Items:**[1]

1. In general, I am among the last in my circle of friends to buy a new _____ when it appears. (r)
2. If I heard that a new _____ was available in the store, I would not be interested enough to buy it. (r)
3. Compared to my friends I own few _____. (r)
4. In general, I am the last in my circle of friends to know the latest _____. (r)
5. I will buy a new _____, even if I haven't heard it yet.
6. I know the names of new _____ before other people do.

---

[1] The name of the product category should be placed in the blank and may require some adjustment in phrasing to be grammatically correct and clear.

# INNOVATIVENESS OF THE CONSUMER (PRODUCT TRIAL)

**Scale Description:**

Eight Likert-type statements are used to measure a consumer's belief that he/she is among the first to try and/or buy new products when they become available. This is in contrast to wanting to stick with previous choices and being reluctant to change. The scale was called *dispositional innovativeness* by Steenkamp and Gielens (2003) as well as Hoffman, Kopalle, and Novak (2010). The name used by Lam et al. (2010) was *consumer innate innovativeness*.

**Scale Origin:**

Steenkamp and Gielens (2003) described the scale as being a revision of the scale used by Steenkamp et al. (1999). That scale was composed of five unspecified items from the Exploratory Acquisition of Products scale (Baumgartner and Steenkamp 1996). In turn, that scale was heavily based on content from the Exploratory Consumer Tendencies scale by Raju (1980).

**Scale Reliability:**

Alphas of .87 (Steenkamp and Gielens 2003), .82 (Hoffman, Kopalle, and Novak 2010), and .79 (Lam et al. 2010) have been reported for the scale.

**Scale Validity:**

The analyses conducted by Steenkamp and Gielens (2003) of this scale and two others provided evidence in support of each scales' unidimensionality as well as their convergent and discriminant validities.

A variety of evidence was provided by Hoffman, Kopalle, and Novak (2010) in support of the scale's unidimensionality and discriminant validity.

Lam et al. (2010) stated that they used EFA and then CFA with all of their reflective measures and they exhibited "good psychometric properties" (p. 138). Information specific to the innovativeness scale was not provided. However, the AVE for the scale was stated to be .53.

## References:

Baumgartner, Hans and Jan-Benedict E. M. Steenkamp (1996), "Exploratory Consumer Buying Behavior: Conceptualization and Measurement," *International Journal of Research in Marketing*, 13 (2), 121-137.

Hoffman, Donna L., Praveen K. Kopalle, and Thomas P. Novak (2010), "The 'Right' Consumers for Better Concepts: Identifying Consumers High in Emergent Nature to Develop New Product Concepts," *Journal of Marketing Research*, 47 (5), 854-865.

Lam, Son K., Michael Ahearne, Ye Hu, and Niels Schillewaert (2010), "Resistance to Brand Switching When a Radically New Brand is Introduced: A Social Identity Theory Perspective," *Journal of Marketing*, 74 (6), 128-146.

Raju, Sekar, H. Rao Unnava, and Nicole Votolato Montgomery (2009), "The Moderating Effect of Brand Commitment on the Evaluation of Competitive Brands," *Journal of Advertising*, 38 (2), 21-35.

Steenkamp, Jan-Benedict E.M. and Katrijn Gielens (2003), "Consumer and Market Drivers of the Trial Probability of New Consumer Packaged Goods," *Journal of Consumer Research*, 30 (December), 368-384.

## Scale Items:[1]

1. When I see a new product on the shelf, I'm reluctant to give it a try. (r)
2. In general, I am among the first to buy new products when they appear on the market.
3. If I like a brand, I rarely switch from it just to try something new. (r)
4. I am very cautious in trying new and different products. (r)
5. I am usually among the first to try new brands.
6. I rarely buy brands about which I am uncertain how they will perform. (r)
7. I enjoy taking chances in buying new products.
8. I do not like to buy a new product before other people do. (r)

[1] A five-point response format was used by Steenkamp and Gielens (2003) while Lam et al. (2010) as well as Hoffman, Kopalle, and Novak (2010) used seven-point scales.

# INNOVATIVENESS OF THE CONSUMER (TECHNOLOGICAL)

**Scale Description:**

Five, seven-point Likert-type statements compose the scale and are intended to measure the degree to which a consumer is motivated to be the first to adopt new technology-based goods and services.

**Scale Origin:**

The scale was developed in a series of studies, some of the details of which are described in an article by Bruner and Kumar (2007a) but with the most details being provided in another publication (Bruner and Kumar 2007b). The version of the scale referred to in the article (2007a) is a subset of items from a larger set developed by the authors for use by *Sprint* to classify customers based on their technological innovativeness (2007b).

**Scale Reliability:**

Based on the studies described by Bruner and Kumar (2007a), the scale had alphas of .91 (Study 1) and .92 (Study 2) using large, national samples.

**Scale Validity:**

Some information bearing on the scale's validity is provided in the article by Bruner and Kumar (2007a) as it was used to help validate another scale they were developing (gadget loving). In particular, evidence was provided in support of the scale's convergent and discriminant validities. Its AVE was .67 and .69 in Studies 1 and 2, respectively.

**References:**

Bruner II, Gordon C., Anand Kumar (2007a), "Gadget Lovers," *Journal of the Academy of Marketing Science*, 35 (3), 329-339.

Bruner II, Gordon C., Anand Kumar, and Clyde Heppner (2007b), "Predicting Innovativeness: Development of the Technology Acceptance Scale," *New Research on Wireless Communications*, Nova Science Publishers, Inc., 1-20.

**Scale Items:**

1. I get a kick out of buying new high tech items before most other people know they exist.
2. It is cool to be the first to own new high tech products.
3. I get a thrill out of being the first to purchase a high technology item.
4. Being the first to buy new technological devices is very important to me.
5. I want to own the newest technological products.

# INNOVATIVENESS OF THE CONSUMER (USAGE)

**Scale Description:**

The scale is composed of five, five-point statements that attempt to capture a consumer's motivation to explore different ways of using a product. Although the product examined by Shih and Venkatesh (2004) was a computer, the statements might be usable with other product categories as well.

**Scale Origin:**

The scale used by Shih and Venkatesh (2004) was heavily based on the Use Innovativeness Index by Price and Ridgway (1983). There are two key differences. First, the full Index had 44 items measuring five factors whereas this scale is sort of a general factor. Shih and Venkatesh (2004) say that they took items from four factors. Second, the Index was not specific to any product whereas the items in this scale refer to a specific product (e.g., computers).

**Scale Reliability:**

An alpha of .81 was reported for the scale by Shih and Venkatesh (2004)

**Scale Validity:**

Shih and Venkatesh (2004) did not provide any information regarding the scale's validity.

**References:**

Price, Linda L. and Nancy M. Ridgway (1983), "Development of a Scale to Measure Use Innovativeness," in *Advances in Consumer Research*, Vol. 10, Richard P. Bagozzi and Alice M. Tybout, eds. Ann Arbor, Michigan: Association for Consumer Research, 679-684.

Shih, Chuan-Fong and Alladi Venkatesh (2004), "Beyond Adoption: Development and Application of a Use-Diffusion Model," *Journal of Marketing*, 68 (January), 59-72.

**Scale Items:**[1]

1. I am creative with _____.
2. I am very curious about how _____ work.
3. I am comfortable working on _____ projects that are different from what I am used to.
4. I often try to do projects on my _____ without exact directions.
5. I use a _____ in more ways than most people do.

---

[1] The extreme verbal anchors for the response scale used by Shih and Venkatesh (2004) were *not at all* (1) and *very much* (5). The name of a product category should be placed in the blanks, e.g., computers.

# INVOLVEMENT WITH TECHNOLOGY

**Scale Description:**

The scale uses three, seven-point Likert-type items to measure the degree to which a person expresses interest in technology and the desire to have new tech products before others.

**Scale Origin:**

The scale was developed by Kim, Haley, and Koo (2009). Along with the other scales used in the main study, this scale was tested in a pretest and revised.

**Scale Reliability:**

Kim, Haley, and Koo (2009) reported the scale to have an alpha of .77 (n = 400 college students).

**Scale Validity:**

A variety of evidence was provided by Kim, Haley, and Koo (2009) in support of the scale's unidimensionality and validity (convergent and discriminant). Its AVE was .58 and .50 for product and corporate ads, respectively.

**Reference:**

Kim, Sora, Eric Haley, and Gi-Yong Koo (2009), "Comparison of the Paths From Consumer Involvement Types To Ad Responses Between Corporate Advertising And Product Advertising," *Journal of Advertising*, 38 (3), 67-80.

**Scale Items:**[1]

1. Technology appeals to me.
2. I am very savvy in _____ technology.

3. When a new technology product comes out, I tend to buy it and try it earlier than others.

---

[1] The name for a particular type of technology can be placed in the blank if desired; e.g., computer.

# LEAD USER (DOMAIN SPECIFIC)

**Scale Description:**

With five, seven-point Likert-type items, the scale measures a person's interest in as well as generation and promotion of new and different ways to satisfy needs within some domain (e.g., product category).

**Scale Origin:**

Hoffman, Kopalle, and Novak (2010) developed and validated the scale along with a companion scale (emergent nature) in an impressive series of studies. The authors' drew upon the work of Morrison, Roberts, and Von Hippel (2000) in conceptualizing the scale. The bulk of the psychometric information regarding the scale came from Study 1. Analysis of its data was based on responses from 1124 adults living in 21 different countries whose native language was English. (See the article and the web appendix for more details of the studies, samples, and scales.)

**Scale Reliability:**

In Study 1, the scale's alpha was .93 (Hoffman, Kopalle, and Novak 2010).

**Scale Validity:**

The Study 1 survey instrument had eight items intended to measure lead user status. Based on the results of an EFA, the scale was reduced to five items. That set, along with items for two other scales, were examined with CFA which provided further evidence of the scale's unidimensionality and discriminant validity.

**Comments:**

With data from the Study 1 sample, the scale's mean was calculated to be 14.61 and the standard deviation was 7.72 (Hoffman, Kopalle, and Novak 2010, web appendix).

## References:

Hoffman, Donna L., Praveen K. Kopalle, and Thomas P. Novak (2010), "The 'Right' Consumers for Better Concepts: Identifying Consumers High in Emergent Nature to Develop New Product Concepts," *Journal of Marketing Research*, 47 (5), 854-865.

Morrison, Pamela D., John H. Roberts, and Eric von Hippel (2000), "Determinants of User Innovation and Innovation Sharing in a Local Market," *Management Science*, 46 (12), 1513–27.

## Scale Items:[1]

1.  Other people consider me as "leading edge" with respect to _____.
2.  I have pioneered some new and different ways for _____.
3.  I have suggested to _____ some new and different ways to _____.
4.  I have participated in offers by _____ to _____ in new and different ways.
5.  I have come up with some new and different solutions to meet my needs for _____.

---

[1] A name for the innovative product or a brief description of it should be inserted in place of the blanks at the end of the items. The phrases used by Hoffman, Kopalle, and Novak (2010) were variations of "home delivery of goods." The short blanks in items #3 and #4 should describe business entities, e.g., stores, companies, online retailers.

# NOVELTY (GENERAL)

**Scale Description:**

The extent to which a person believes that something is uncommon and distinct is measured in this scale with four, uni-polar items along with a seven-point Likert-type response format. The scale is general in the sense that the items are amenable for use in a wide variety of situations when participants are given the proper instructions.

**Scale Origin:**

The scale used by Yim, Cicchirillo, and Drumwright (2012) is based on items found in several measures of novelty used in previous research. In particular, the authors cited Massetti (1996) and, indeed, it seems to be most similar to it. Yim, Cicchirillo, and Drumwright (2012) used the scale in both studies discussed in their article. Study 1 was composed of 85 undergraduate students at a U.S. university while the sample in Study 2 was 108 students at a different U.S. university.

**Scale Reliability:**

The alpha for the scale was .93 which was based, apparently, on the combined samples from both studies (Yim 2013).

**Scale Validity:**

Yim, Cicchirillo, and Drumwright (2012) used CFA to assess the psychometric quality of several, if not all, of their scales. The implication was that support was found for the convergent and discriminant validities of the scales. However, no specific statistics were provided for the novelty scale and it is not clear that it was included in the CFA.

## References:

Massetti, Brenda (1996), "An Empirical Examination of the Value of Creativity Support Systems on Idea Generation," *MIS Quarterly*, 20 (1), 83–97.

Yim, Mark Yi-Cheon (2013), personal correspondence.

Yim, Mark Yi-Cheon, Vincent J. Cicchirillo, and Minette E. Drumwright (2012), "The Impact of Stereoscopic Three-Dimensional (3-D) Advertising," *Journal of Advertising*, 41 (2), 113-128.

## Scale Items:[1]

1. new
2. unique
3. different
4. unusual

---

[1] The extreme verbal anchors used with the items were *strongly disagree* and *strongly agree* (Yim, Cicchirillo, and Drumwright 2012, p. 118).

# PACE OF TECHNOLOGICAL INNOVATIVENESS

**Scale Description:**

The scale is composed of six, seven-point Likert-type items intended to measure the extent to which a person views the rate of technological change in a particular product category to be high.

**Scale Origin:**

The scale was developed by Grewal, Mehta, and Kardes (2004).

**Scale Reliability:**

Grewal, Mehta, and Kardes (2004) reported a construct reliability of .97 for the scale.

**Scale Validity:**

Using confirmatory factor analysis, Grewal, Mehta, and Kardes (2004) examined a measurement model of this scale and two others (necessity-luxury product character and public-private product character). The analysis provided evidence in support of each scale's convergent and discriminant validities.

**Reference:**

Grewal, Rajdeep, Raj Mehta, Frank R. Kardes (2004), "The Timing of Repeat Purchases of Consumer Durable Goods: The Role of Functional Bases of Consumer Attitudes," *Journal of Marketing Research*, 41 (February), 101-115.

**Scale Items:**[1]

In your opinion, how would you rate the pace of technological innovation in _____ ?

1. _____ technology is changing at a very fast pace.

2. Compared to other consumer durable products, _____ technology is changing fast.

3. I have NOT seen significantly new technology in _____ for some time. (r)

4. Innovations in _____ are very frequent.

5. Pace of technological innovations in _____ is high.

6. Technological innovations and _____ don't go hand in hand. (r)

---

[1] The name of the product should be placed in the blanks, e.g., car(s).

# RELATIVE ADVANTAGE OF THE PRODUCT

**Scale Description:**

The scale is composed of three, seven-point Likert-type statements that measure the degree to which a consumer believes that a particular good or service performs some function better than competing products. Because this is one of the five key characteristics that are thought to influence adoption of innovations (Rogers 2003), the construct is most typically examined with respect to new products rather than mature ones.

**Scale Origin:**

The scale is original to Meuter et al. (2005) but they drew some key phrases and concepts from a scale by Moore and Benbasat (1991).

**Scale Reliability:**

An alpha of .95 was reported by Meuter et al. (2005) for use of the scale in both of their studies.

**Scale Validity:**

At a general level, Meuter et al. (2005) tested a measurement model containing all of their constructs and indicators. Its fit was acceptable. The factor loadings were reported to be significant and evidence of discriminant validity was provided for each construct using two different tests (confidence interval, variance extracted).

**References:**

Meuter, Matthew L., Mary Jo Bitner, Amy L. Ostrom, and Stephen W. Brown (2005), "Choosing Among Alternative Service Delivery Modes: An Investigation of Customer Trial of Self-Service Technologies," *Journal of Marketing*, 69 (April), 61-83.

Moore, Gary C. and Izak Benbasat (1991), "Development of an Instrument to Measure the Perceptions of Adopting an Information Technology Innovation," *Information Systems Research*, 2 (3), 192-223.

Rogers, Everett M. (2003), *Diffusion of Innovations*, New York: The Free Press.

**Scale Items:**[1]

1. Using the _____ improves the _____.
2. Overall, I believe using the _____ is advantageous.
3. I believe the _____, in general, is the best way to _____.

---

[1] The name of the good or service should be placed in the blank of #2 and the first blanks of #1 and #3. The second blanks of #1 and #3 should be filled with a brief description of the product's function, e.g., order a prescription refill (Meuter et al. 2005).

# TECHNOLOGICAL ANXIETY

**Scale Description:**

Four statements are used to measure the degree to which a consumer is apprehensive about technology and avoids its usage. This construct is sometimes referred to by the more provocative term *technophobia* (e.g., Brosnan 1998; Rosen, Sears, and Weil 1987).

**Scale Origin:**

Meuter et al. (2005) cited Raub (1981) as the source from which they adapted items. Keh and Pang (2010) as well as Lakshmanan and Krishnan (2011) cited Meuter et al. (2005) as the developers.

With respect to the study by Keh and Pang (2010), data were gather in China. It is not clear if the scale was presented in Chinese or English.

**Scale Reliability:**

An alpha of .93 was reported by Meuter et al. (2005) for use of the scale in both of their studies. Keh and Pang (2010) used the scale in Study 3 and its alpha was .896. The scale was used in three of the four studies described in the article by Lakshmanan and Krishnan (2011) but the alpha was only reported for Study 1 where it was .85.

**Scale Validity:**

At a general level, Meuter et al. (2005) tested a measurement model containing all of their constructs and indicators. Its fit was acceptable. The factor loadings were reported to be significant and evidence of discriminant validity was provided for each construct using two different tests (confidence interval, variance extracted).

Keh and Pang (2010) used CFA to examine several scales in Study 3. Evidence was provided in support of each scale's unidimensionality as well their convergent and discriminant validities.

Lakshmanan and Krishnan (2011) did not discuss the scale's validity.

## References:

Brosnan, Mark J. (1998), *Technophobia: The Psychological Impact of Information Technology*, London: Routledge.

Keh, Hean Tat and Jun Pang (2010), "Customer Reactions to Service Separation," *Journal of Marketing*, 74 (2), 55-70.

Lakshmanan, Arun and H. Shanker Krishnan (2011), "The Aha! Experience: Insight and Discontinuous Learning in Product Usage," *Journal of Marketing*, 75 (6), 105-123.

Meuter, Matthew L., Mary Jo Bitner, Amy L. Ostrom, and Stephen W. Brown (2005), "Choosing Among Alternative Service Delivery Modes: An Investigation of Customer Trial of Self-Service Technologies," *Journal of Marketing*, 69 (April), 61-83.

Raub, Annalyse Callahan (1981), *Correlates of Computer Anxiety in College Students*, doctoral dissertation, University of Pennsylvania.

Rosen, Larry D., Deborah C. Sears, and Michelle M. Weil (1993), "Treating Technophobia: A Longitudinal Evaluation of the Computerphobia Reduction Program," *Computers in Human Behavior*, 9 (1), 27-50.

## Scale Items:[1]

1. I feel apprehensive about using technology.
2. Technical terms sound like confusing jargon to me.
3. I have avoided technology because it is unfamiliar to me.
4. I hesitate to use most forms of technology for fear of making mistakes I cannot correct.

---

[1] Meuter et al. (2005) as well as Lakshmanan and Krishnan (2011) used a seven-point Likert-type response format with these items. The format used by Keh and Pang (2010) was not clearly described but appears to have also been seven-point, Likert-type.

# TECHNOLOGY USAGE DISCOMFORT

**Scale Description:**

Four statements along with a seven-point Likert-type response format are used in this scale to measure the degree to which a person generally has difficulty understanding and using technological goods and services.

**Scale Origin:**

The scale used by Zhu et al. (2007) is a subset of items from the Technology Readiness instrument by Parasuraman (2000). Zhu et al. (2007) selected items that expressed a person's lack of expertise and confidence with regard to using technology.

**Scale Reliability:**

Zhu et al. (2007) reported the alpha for the scale to be .68 (Experiment 2). The internal consistency for this scale is rather low and may be due to one or more items not tapping into the same construct as well as the others do.

**Scale Validity:**

Zhu et al. (2007) did not address the validity of this specific scale but they did note that all items in their study loaded on their expected constructs in a CFA and the model fit for the constructs was satisfactory.

**References:**

Parasuraman, A. (2000), "Technology Readiness Index (TRI)," *Journal of Services Marketing*, 2 (May), 307-320.

Zhu, Zhen, Cheryl Nakata, K. Sivakumar, and Dhruv Grewal (2007), "Self-Service Technology Effectiveness: the Role of Design Features and Individual Traits," *Journal of the Academy of Marketing Science*, 35 (4), 492-506.

**Scale Items:**

1. When I get technical support from a company of a high-tech product or service, I sometimes feel as if I am being taken advantage of by someone who knows more than I do.
2. If I buy a high-tech product or service, I prefer to have the basic model over one with a lot of extra features.
3. It is embarrassing when I have trouble with a high-tech gadget while people are watching.
4. Technology always seems to fail at the worst possible time.

# TECHNOLOGY USAGE MOTIVATION (INTRINSIC)

**Scale Description:**

Four, seven-point Likert-type statements are used to measure the degree to which a consumer believes that use of a certain piece of technology would lead to positive, personal consequences (enjoyment, independence, confidence). The scale was called *perceived value in future co-creation* by Dong, Evans, and Zou (2008) and was the *instrumentality* dimension of intrinsic motivation in the study by Meuter et al. (2005).

**Scale Origin:**

The four item scale used by Dong, Evans, and Zou (2008) is taken from a five item scale by Meuter et al. (2005).

**Scale Reliability:**

An alpha of .88 was reported by Dong, Evans, and Zou (2008).

**Scale Validity:**

Dong, Evans, and Zou (2008) tested a measurement model containing all of their constructs' indicators and its fit was acceptable. The authors' stated that evidence was found in support of this scale's convergent and discriminant validities.

**References:**

Dong, Beibei, Kenneth R. Evans, and Shaoming Zou (2008), "The Effects of Customer Participation in Co-Created Service Recovery, *Journal of the Academy of Marketing Science*, 36 (1), 123-137.

Meuter, Matthew L., Mary Jo Bitner, Amy L. Ostrom, and Stephen W. Brown (2005), "Choosing Among Alternative Service Delivery Modes: An Investigation of Customer Trial of Self-Service Technologies," *Journal of Marketing*, 69 (April), 61-83.

**Scale Items:**

Using the _____:[1]

1. would provide me with personal feelings of worthwhile accomplishment.
2. would provide me with feelings of enjoyment from using the technology.
3. would provide me with feelings of independence.
4. would allow me to feel innovative in how I interact with a service provider.
5. would allow me to have increased confidence in my skills.

---

[1] The name of the technology good or service should be placed in the blanks.

# USEFULNESS (GENERAL)

**Scale Description:**

Eight, seven-point semantic-differentials are used to measure the degree of functional value a person believes a particular object (product, process, etc.) has. Coupled with the proper instructions, the scale items are amenable for use in a wide variety of contexts.

**Scale Origin:**

The scale was developed by Kleijnen, de Ruyter, and Wetzels (2007) who built upon a scale by Voss, Spangenberg, and Grohmann (2003). Kleijnen, de Ruyter, and Wetzels (2007) pretested the scale along with other scales in preparation for their main study.

**Scale Reliability:**

Kleijnen, de Ruyter, and Wetzels (2007) used the scale with respect to three "channels" in which financial transactions could take place. The alphas were .95 (mobile), .97 (non-mobile electronic), and .95 (retail).

**Scale Validity:**

Besides some facets of validity examined in their pretest, Kleijnen, de Ruyter, and Wetzels (2007) provided support for the scale's convergent and discriminant validities in the main study. The scale's AVEs were .69 (mobile), .81 (non-mobile electronic), and .72 (retail).

**References:**

Kleijnen, Mirella, Ko de Ruyter, and Martin Wetzels (2007), "An Assessment of Value Creation in Mobile Service Delivery and the Moderating Role of Time Consciousness," *Journal of Retailing*, 83 (1), 33-46.

Voss, Kevin E., Eric R. Spangenberg, and Bianca Grohmann (2003), "Measuring the Hedonic and Utilitarian Dimensions of Consumer Attitude," *Journal of Marketing Research*, 40 (August) 310–320.

**Scale Items:**

1.  Ineffective / effective
2.  Not functional / functional
3.  Impractical / practical
4.  Useless / useful
5.  Not sensible / sensible
6.  Inefficient / efficient
7.  Unproductive / productive
8.  Bad / good

# USEFULNESS OF THE PRODUCT

**Scale Description:**

Using three, five-point Likert-type items, this scale measures the quality of a product with the focus on the improvement it makes in one's productivity.  The scale is best suited for an innovation which has benefits of a functional nature as opposed to hedonic or social.

**Scale Origin:**

Van Ittersum and Feinberg (2010) did not identify the source of the scale except to say they "relied on published, validated scales" in three of their studies.  The authors called the scale *perceived relative advantage,* referring to the construct long identified by Rogers (e.g., 2003) as one of the key characteristics of successful innovations.  Yet, the authors admitted that their operationalization was more absolute than relative since the items do not have the respondent compare the advantage(s) of a focal innovation to alternatives.

In Study 1, 212 superintendents of U.S. golf courses were surveyed about their intentions to adopt an advanced grass mower.  In Study 2, 266 U.S. farm operators were asked about their intention to adopt an autoguidance system for their tractors.  Finally, 354 U.S. college students were asked in Study 3 about their intentions to adopt cell phones with GPS.

**Scale Reliability:**

Alphas for the scale were .81, .88., and .93 in Studies 1, 2, and 3, respectively (Van Ittersum and Feinberg 2010).

**Scale Validity:**

Van Ittersum and Feinberg (2010) did not address the scale's validity.

## References:

Rogers, Everett M. (2003), *Diffusion of Innovations*, New York: The Free Press.

Van Ittersum, Koert and Fred M. Feinberg (2010), "Cumulative Timed Intent: A New Predictive Tool for Technology Adoption," *Journal of Marketing Research*, 47 (5), 808-822.

## Scale Items:[1]

1. Using _____ in my life/work would increase my productivity.
2. If I use _____, I will increase the quality of output.
3. Using _____ increases my productivity.

---

[1] The product or brand name should be placed in the blanks.

# VISIBILITY OF THE NEW PRODUCT'S USAGE

**Scale Description:**

This four-item, seven-point Likert-type scale is used for measuring the perception that if a new product were purchased it would be noticed by a social group important to the consumer.

**Scale Origin:**

Although not specifically stated by Fisher and Price (1992), the scale is original (Fisher 1994). Items composing this scale were refined along with items intended to measure four other constructs. No information about this scale in particular was provided. In general, item-to-total correlations had to be more than .50 and items had to load on hypothesized factors. Items that did not fit these criteria were eliminated before use in the main study.

**Scale Reliability:**

Fisher and Price (1992) reported an alpha of .86 for the scale.

**Scale Validity:**

Fisher and Price (1992) did not specifically address the validity of the scale. However, they did state that the variance extracted for the scale was .67, which provides some limited evidence of its convergent validity.

**Comments:**

Fisher and Price (1992) used the scale to measure students' perceptions of a new product idea: cordless headphones. It seems possible that the statements could be adapted for use with other products and types of reference groups.

## References:

Fisher, Robert J. (1994), personal correspondence.

Fisher, Robert J. and Linda L. Price (1992), "An Investigation Into the Social Context of Early Adoption Behavior," *Journal of Consumer Research,* 19 (December), 477-86.

## Scale Items:[1]

1. If I bought one of the _____, other students would see me using it.
2. If I were to buy a _____, other students would recognize that I owned something that is new to the market.
3. It's likely that other students would notice if I were to buy a

   _____.

4. If I bought a _____, other students would see that I had something that is unusual.

---

[1] The items were supplied by Fisher (1994). The generic name of the product should be placed in the blanks. The term "students" should be replaced with a term/phrase that describes an important reference group of which the respondent is a member (e.g., shoppers, employees, neighbors). If a general term such as "persons" is used, it changes the meaning of the scale somewhat.

## ABOUT THE AUTHOR

Dr. Bruner (PhD, University of North Texas) has spent over 30 years as a university professor with Consumer Behavior being his specialty. In terms of publications, he is best known for the *Marketing Scales Handbook* series which since 1992 has described and reviewed thousands of multi-item measures for use by researchers around the world in academia and industry. Additionally, Dr. Bruner has considerable experience studying technology-related issues. Given these experiences, he is uniquely qualified to produce this book which provides researchers with 30 measures that are especially relevant when studying consumers' thoughts and behaviors regarding innovations, particularly those of a high-tech nature.